The Canadian Oil Sands
Energy Security vs. Climate Change

COUNCIL on
FOREIGN
RELATIONS

Center for Geoeconomic Studies

Council Special Report No. 47
May 2009

Michael A. Levi

The Canadian Oil Sands
Energy Security vs. Climate Change

The Council on Foreign Relations (CFR) is an independent, nonpartisan membership organization, think tank, and publisher dedicated to being a resource for its members, government officials, business executives, journalists, educators and students, civic and religious leaders, and other interested citizens in order to help them better understand the world and the foreign policy choices facing the United States and other countries. Founded in 1921, CFR carries out its mission by maintaining a diverse membership, with special programs to promote interest and develop expertise in the next generation of foreign policy leaders; convening meetings at its headquarters in New York and in Washington, DC, and other cities where senior government officials, members of Congress, global leaders, and prominent thinkers come together with CFR members to discuss and debate major international issues; supporting a Studies Program that fosters independent research, enabling CFR scholars to produce articles, reports, and books and hold roundtables that analyze foreign policy issues and make concrete policy recommendations; publishing *Foreign Affairs*, the preeminent journal on international affairs and U.S. foreign policy; sponsoring Independent Task Forces that produce reports with both findings and policy prescriptions on the most important foreign policy topics; and providing up-to-date information and analysis about world events and American foreign policy on its website, www.cfr.org.

The Council on Foreign Relations takes no institutional position on policy issues and has no affiliation with the U.S. government. All statements of fact and expressions of opinion contained in its publications are the sole responsibility of the author or authors.

Council Special Reports (CSRs) are concise policy briefs, produced to provide a rapid response to a developing crisis or contribute to the public's understanding of current policy dilemmas. CSRs are written by individual authors—who may be CFR fellows or acknowledged experts from outside the institution—in consultation with an advisory committee, and are intended to take sixty days from inception to publication. The committee serves as a sounding board and provides feedback on a draft report. It usually meets twice—once before a draft is written and once again when there is a draft for review; however, advisory committee members, unlike Task Force members, are not asked to sign off on the report or to otherwise endorse it. Once published, CSRs are posted on www.cfr.org.

For further information about CFR or this Special Report, please write to the Council on Foreign Relations, Communications, 58 East 68th Street, New York, NY 10065, or call the Communications office at 212.434.9888. Visit our website, www.cfr.org.

Copyright © 2009 by the Council on Foreign Relations® Inc.
All rights reserved.
Printed in the United States of America.

This report may not be reproduced in whole or in part, in any form beyond the reproduction permitted by Sections 107 and 108 of the U.S. Copyright Law Act (17 U.S.C. Sections 107 and 108) and excerpts by reviewers for the public press, without express written permission from the Council on Foreign Relations. For information, write to the Publications Office, Council on Foreign Relations, 58 East 68th Street, New York, NY 10065.

To submit a letter in response to a Council Special Report for publication on our website, www.cfr.org, you may send an email to CSReditor@cfr.org. Alternatively, letters may be mailed to us at: Publications Department, Council on Foreign Relations, 58 East 68th Street, New York, NY 10065. Letters should include the writer's name, postal address, and daytime phone number. Letters may be edited for length and clarity, and may be published online. Please do not send attachments. All letters become the property of the Council on Foreign Relations and will not be returned. We regret that, owing to the volume of correspondence, we cannot respond to every letter.

Cover Photo: A dump truck carries a 400-ton load of oil sands ore in a mining operation in Fort McMurray, Alberta, Canada (Larry Macdougal/Getty Images).

This report is printed on paper certified by SmartWood to the standards of the Forest Stewardship Council, which promotes environmentally responsible, socially beneficial, and economically viable management of the world's forests.

Contents

Foreword vii
Acknowledgments ix
Acronyms xi

Council Special Report 1
Introduction 3
Status, Prospects, and Challenges 5
Energy Security and Climate Change 15
Policy Recommendations 28

Endnotes 41
About the Author 44
Advisory Committee 45
CGS Advisory Committee 46
CGS Mission Statement 48

Foreword

Rhetoric in Washington often focuses on areas where energy security and climate change, two increasingly prominent elements of U.S. domestic and foreign policy, align. Many important decisions, though, will require difficult trade-offs between them. The Canadian oil sands —a massive but emissions-intensive source of oil—presents policymakers with precisely such a challenge. Unfettered production in the oil sands would increase greenhouse gas emissions but strengthen U.S. energy security with a supply of oil from a friendly and stable neighbor. Sharply curtailed oil sands operations would harm U.S. energy security but cut emissions.

This Council Special Report, authored by Michael A. Levi, explores both the energy security and climate change implications of expanded oil sands production. It assesses current and future trends in the oil sands, including in the scale and cost of production and in the oil sands' impact on world oil markets. The report concludes that the oil sands are neither critical to U.S. energy security nor catastrophic for climate change. It also argues, though, that their security benefits and climate costs cannot be ignored. The report's recommendations focus on policies that would provide incentives to cut the emissions generated in producing each barrel of crude from the oil sands, but in a way that is careful to avoid directly discouraging increased production. The recommended measures do not fully satisfy narrow energy security or climate change concerns, but instead seek to balance them.

The Canadian Oil Sands: Energy Security vs. Climate Change makes an important contribution on a subject that will be central to energy and climate debates. Canadian policymakers and global oil markets will directly shape the oil sands' development, but because the United States is the natural destination for many oil sands products, U.S.

decisions will inevitably play a critical role. This report offers a nuanced and thoughtful examination of the relevant issues and of options for U.S. policy.

Richard N. Haass
President
Council on Foreign Relations
May 2009

Acknowledgments

I am most grateful to the members of this report's advisory committee, and in particular to Ernest J. Moniz for chairing the group. The advisory group's members' spirited debates and individual advice were invaluable.

In developing this Council Special Report, I also interviewed a variety of individuals who were supportive of, ambivalent toward, and opposed to oil sands development. These included individuals from governments, industry, and nongovernmental organizations. I thank them all for sharing their time and insights.

I am thankful to my research associate, Katherine Michonski, who skillfully supported my research and writing, and provided consistently thoughtful feedback as I was developing my ideas and arguments.

I thank CFR President Richard N. Haass for his comments on a draft of this report. Thank you also to Patricia Dorff and Lia Norton in Publications, and to Lisa Shields and Anya Schmemann in Communications and Marketing, for their efforts in producing and disseminating this study.

This report was made possible by a grant from the Rockefeller Foundation and through David M. Rubenstein's generous support of CFR's work on the critical international challenges involving energy and the environment. The statements made and views expressed in this report are solely my responsibility.

Michael A. Levi

Acronyms

ANWR	Arctic National Wildlife Refuge
CCS	carbon capture and sequestration
CERI	Canadian Energy Research Institute
EIA	Energy Information Administration
IEA	International Energy Agency
LCFS	low-carbon fuel standard
LNG	liquefied natural gas
mb/d	million barrels per day
NAFTA	North American Free Trade Agreement
NRDC	Natural Resources Defense Council
OPEC	Organization of Petroleum Exporting Countries
RD&D	research, development, and demonstration
SAGD	steam-assisted gravity drainage
WTI	West Texas Intermediate
WTO	World Trade Organization

Council Special Report

Introduction

Half a decade of high and volatile oil prices alongside increasingly dire warnings of climatic disaster have pushed energy security and climate change steadily up the U.S. policy agenda. Rhetoric in Washington has emphasized opportunities to deal with both challenges at once. But energy security and climate change do not always align: many important decisions in areas including unconventional oil, biofuels, natural gas, coal, and nuclear power will involve complex trade-offs and force policymakers to carefully navigate the two goals. Ongoing and heated debates in the United States and Canada over the future of the Canadian oil sands—touted at once as an energy security godsend and a climate change disaster—highlight that tension and emphasize the need to intelligently address it.

The oil sands (often referred to as tar sands) are largely contained within the Canadian province of Alberta. Policymakers on both sides of the border understand, though, that their development will have security, economic, and environmental ramifications that extend well beyond Canada, and that many U.S. energy and climate decisions will inevitably have major implications for the oil sands' future. U.S. federal and state legislators have already proposed several laws that would affect the oil sands. Canadian policymakers also understand that the United States will play a critical role: Canadian minister of environment Jim Prentice recently gave a speech calling for "a bilateral approach to the environment and to energy" featuring "a common carbon market" and "a level playing field" while highlighting Canada's immense reserves and arguing that Canada "should play a larger role in the North American energy security solution."

This report assesses the energy security and climate change impacts of the oil sands and makes recommendations for U.S. policymakers within the context of broader bilateral relations. The first section reviews projected oil sands production through 2030 and assesses how oil prices and other nonmarket forces are likely to shape the oil sands' development; it also examines how changes in the oil sands might affect world energy markets. Based on that foundation, the second section assesses the likely energy security and climate change impacts of oil sands expansion, leading to a set of principles for balancing the two goals. The third section reviews the current state of Canadian and U.S. policy and then provides recommendations for U.S. policymakers.

Status, Prospects, and Challenges

The Canadian oil sands are a mixture of sand, clay, and bitumen, a highly dense and viscous tar-like form of petroleum. They are concentrated primarily in the Canadian province of Alberta. The International Energy Agency (IEA) estimates that the oil sands contain nearly 1.7 trillion barrels of oil. Proven reserves—those that can be extracted given prevailing and expected economic and operating conditions—were estimated to exceed 170 billion barrels as of January 2008, ranking Canada second only to Saudi Arabia.[1] This is much larger than the resource contained, for example, in the environmentally controversial Arctic National Wildlife Refuge (ANWR), which is estimated to have less than ten billion barrels.[2]

The oil sands yielded 1.2 million barrels a day (mb/d) in 2006, triple their level in 1990.[3] This was equal to 1.4 percent of global oil production and to roughly 6 percent of total U.S. oil consumption, 9 percent of U.S. oil imports (including refined products), and 24 percent of U.S. domestic oil production. Indeed, since 2004 Canada has been the biggest source of U.S. oil imports.

The oil sands' future potential is harder to assess: it depends on global oil prices as well as on the availability of oil worldwide, all of which will be shaped by physical and political conditions that are hard to project. Understanding that future, though, is essential to assessing the oil sands' significance to U.S. energy security and to judging their likely climate impact. This section of the report outlines expected trends in oil sands production, describes how oil prices and nonmarket barriers might affect the cost and volume of future production, and then assesses how changes in oil sands volumes and costs might affect world oil markets.

TRENDS

Figure 1 shows projections for oil sands production released in March 2009 by the U.S. Energy Information Administration (EIA) in its *Annual Energy Outlook*, as well as the EIA's previous projection from June 2008; the oil price assumptions for the four cases in Figure 1 are shown in Figure 2. (These projections, along with the others in this section, assume that governments implement no emissions-constraining policies beyond those that are already in place.) Oil sands production more than triples by 2030 (to 4.3 mb/d) in the reference case. (One can again contrast that with ANWR, where the EIA projects that production would peak at 0.8 mb/d before 2030 and then decline.[4]) The figures are similar to (though more bullish in the short term than) projections published in February 2009 by the Canadian Energy Research Institute (CERI), which forecast that production would rise to about 2.2 mb/d in 2015 and about 4.2 mb/d in 2030; the CERI numbers for 2015, which are about 20 percent lower than the EIA figures, are likely to be more accurate.[5] Figures 3 and 4 show that oil sands are expected to make up a rising share of both global and non-OPEC oil supply, and an even larger fraction of growth in non-OPEC production.

These recent projections are significantly lower than those made in mid-2008, as the ongoing recession has forced analysts to lower their production projections and, more broadly, has introduced much greater uncertainty into forecasting. They are higher than many projections made a year ago, though, as expected, long-term oil prices have increased.

Oil prices have temporarily collapsed as global demand has sunk. That drop, from $120 per barrel as late as October 2008 to a band roughly between $35 and $50 during 2009, has had two effects.[6] It has made new oil sands projects less profitable, leading to cancellations and delays. Lower prices have also meant that many oil companies have less capital available even for theoretically attractive expansion.

COSTS

Longer-term capital and operating cost projections have also been thrown into question. Prices for steel, cement, and natural gas (all critical inputs into oil sands production) have fallen sharply, though many producers are locked into long-term contracts; labor costs, which

FIGURE 1. PROJECTED OIL SANDS PRODUCTION

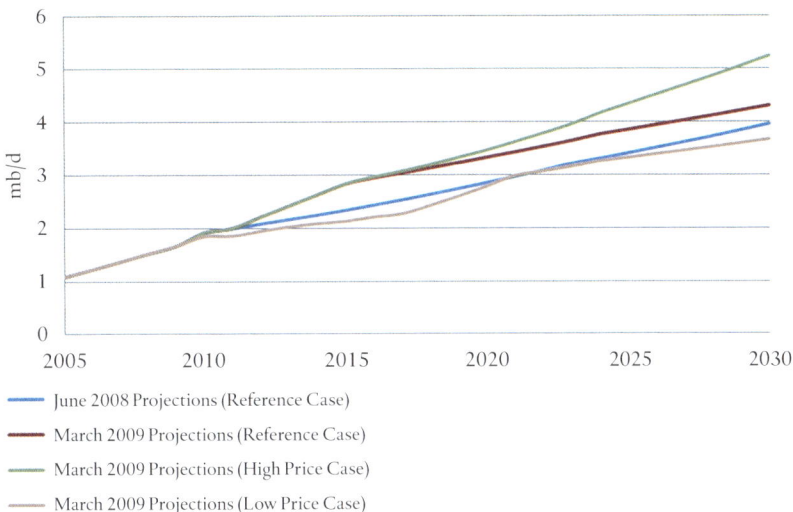

— June 2008 Projections (Reference Case)
— March 2009 Projections (Reference Case)
— March 2009 Projections (High Price Case)
— March 2009 Projections (Low Price Case)

Note: The low price case is indicative of what might happen to oil sands if prices were restrained by global efforts to control consumption; the high price case might describe a world in which cheap oil turns out to be less plentiful than currently expected.

Sources: EIA, *Annual Energy Outlook 2009*; EIA, *Annual Energy Outlook 2008*.

FIGURE 2. ASSUMED PRICE TRAJECTORIES FOR PROJECTIONS IN FIGURE 1

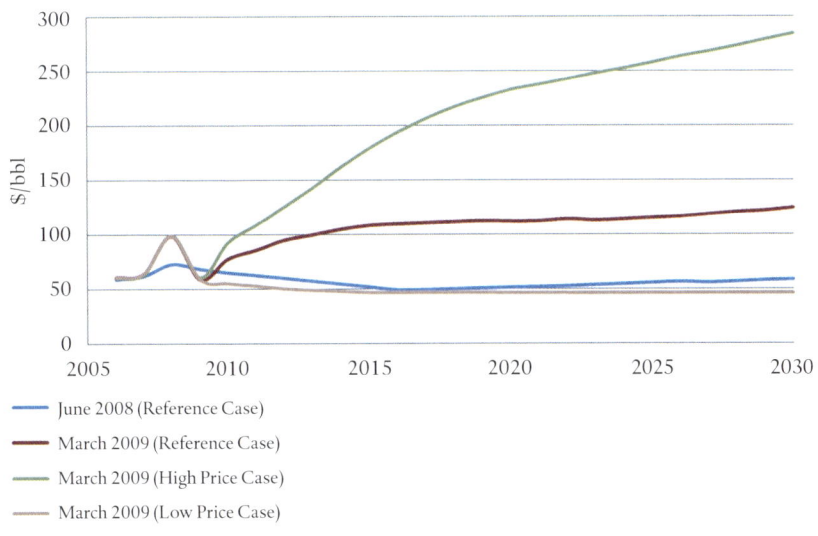

— June 2008 (Reference Case)
— March 2009 (Reference Case)
— March 2009 (High Price Case)
— March 2009 (Low Price Case)

Sources: EIA, *Annual Energy Outlook 2009*; EIA, *Annual Energy Outlook 2008*.

FIGURE 3. PROJECTED OIL SANDS PRODUCTION AS A PERCENTAGE OF GLOBAL SUPPLY

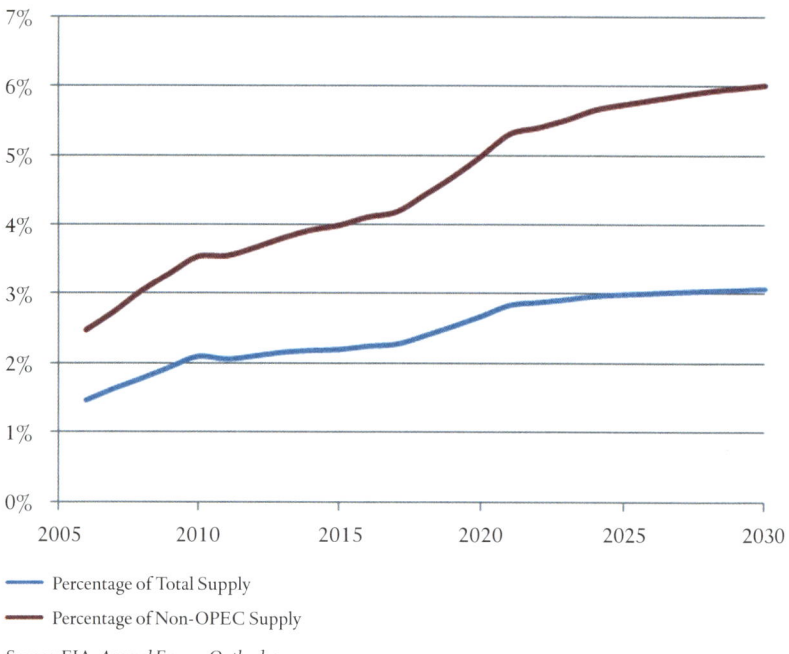

Source: EIA, *Annual Energy Outlook 2009*.

tend to drop more slowly than commodity prices, should decline over time as well. When economic recovery eventually drives the price of oil back up, though, commodity and labor prices can be expected to rise too. How much is difficult to predict, but the details will determine which oil sands projects will be profitable in the future, as well as how various policy measures will affect their viability. Regulatory uncertainty—the Alberta government has revised its royalty and tax structures repeatedly—only adds to the confusion.

There is, nonetheless, an emerging consensus that production from existing oil sands projects will continue to be economically viable at world oil prices exceeding roughly $35 to $40 per barrel. Companies are also expected to continue producing at many existing projects even if prices drop to $30 for several months since it can be expensive and time consuming to restart many operations. This means that near-term production is unlikely to decline.

FIGURE 4. PROJECTED OIL SANDS PRODUCTION GROWTH AS A PERCENTAGE OF TOTAL SUPPLY GROWTH

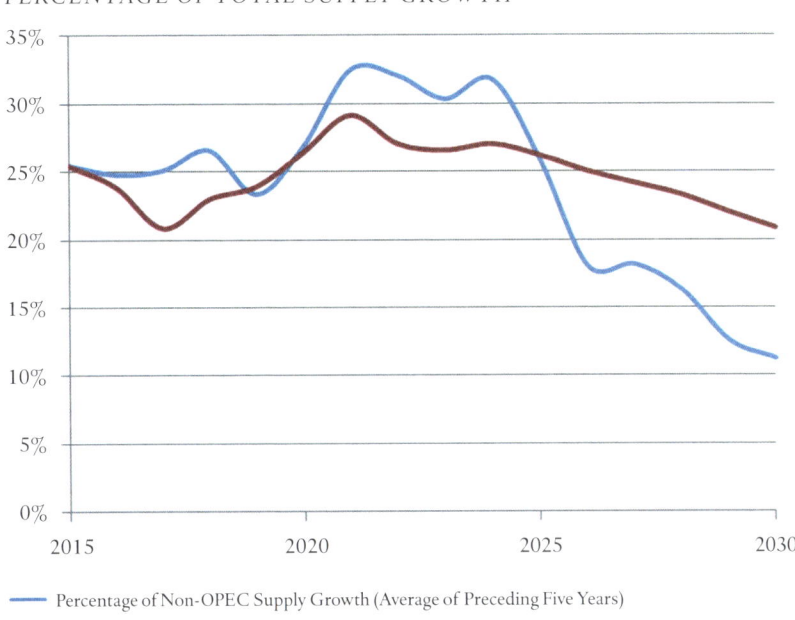

— Percentage of Non-OPEC Supply Growth (Average of Preceding Five Years)
— Percentage of Non-OPEC Supply Growth (Cumulative from 2010)

Note: Figures beyond 2025 are highly speculative.

Source: EIA, *Annual Energy Outlook 2009*.

New projects will need the price of light sweet crude oil, such as West Texas Intermediate (WTI), to be sustained at significantly higher levels in order to be profitable.[7] Analysts frequently point to a threshold in the $60 to $70 per barrel range for new investments; such prices are likely, though not certain, to return in the next few years. The threshold for some in situ projects, which generally have lower initial capital costs, may be as much as $10 per barrel ($10/bbl) lower, while that for some capital-intensive mining projects may be as much as $10/bbl higher. (Oil sands producers fetch lower prices for their heavier products than those that prevail for more desirable crudes; that is already reflected in this estimate.) These figures are substantially greater than the threshold was only a few years ago, when prices of $30 and up were considered sufficient to justify new projects; they are also lower than the threshold for new projects that prevailed in early 2008, when labor and equipment shortages, along with spiking commodity prices, led to

OIL SANDS PRODUCTION

Oil sands production is different from conventional oil production. Two basic approaches are used. Resources within about 100 meters of the surface are mined and then processed in facilities where the bitumen they contain is extracted. Deeper deposits, which comprise about four-fifths of Alberta's resources, must be produced using so-called in situ methods. The most widely used of these is steam-assisted gravity drainage (SAGD): two horizontal wells are drilled; hot steam is pumped into the upper one, causing bitumen to flow into the lower well, from which it is drawn. The bitumen from either mining or in situ operations is then either "upgraded" at a separate plant to synthetic crude oil (SCO) or mixed with other liquids to make "dilbit" or "synbit." Depending on the degree of upgrading, SCO is either processed in refineries designed for light crudes or must be processed in refineries tailored to heavy crudes. Refineries must be specially modified to process dilbit and synbit.

OIL SANDS SUPPLY CHAIN

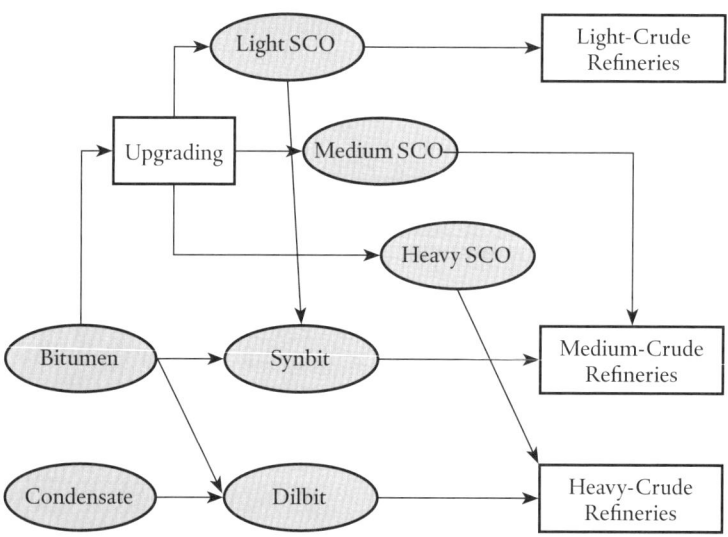

Source: Michael Toman et al., *Unconventional Fossil-Based Fuels: Economic and Environmental Trade-Offs* (Santa Monica, CA: RAND Corporation, 2008). Used with permission.

spiraling costs. The new numbers reflect a belief that some of the price run-up in the last few years was cyclical, but that there have also been structural changes in the markets underlying oil sands development that will permanently increase costs.

NONMARKET BARRIERS

There are several wild cards that could in principle curtail oil sands production. This report reviews three here: natural gas availability, water scarcity, and public opposition due to local social and environmental impacts. (Climate policy constraints are addressed later.) Each is already reflected to some extent in the projections just presented, but each still presents real risks. In particular, even though the production volume estimates just presented are fairly robust, water scarcity and strong public opposition could constrain growth. Ultimately, though, the boost the oil sands provide to the Albertan and Canadian economies gives both governments strong incentives to resolve these challenges in a way that allows robust continued expansion.

Current methods of exploiting the oil sands require large amounts of natural gas; total natural gas purchases for oil sands operations are about 900 million cubic feet per day, or 5 percent of Canada's natural gas production. Assuming rapid expansion of oil sands production, though, natural gas demands have been projected to rise to anywhere between 2.2 billion and 3.2 billion cubic feet per day by 2020.[8] (This trend will be moderated by slower than previously expected oil sands growth.) Canadian natural gas production, meanwhile, is forecast to be similar in 2020 to its current level (though it is expected to temporarily decline in the interim), which would make the draw for oil sands production a much higher fraction.[9] However, the fact that most Canadian gas is, like the oil sands, located in Alberta makes it politically unlikely that the government will artificially restrict gas availability to the oil sands. Operators are also exploring a variety of alternatives to natural gas in order to hedge risk and, in some cases, to cut costs.

Oil sands production is also highly water intensive; water availability might thus limit future operations. The constraint pertains primarily to mining projects, which require between 2 and 4.5 barrels of water for each barrel of oil that they produce; in situ projects are able to recycle water, thus limiting their net use to about 0.2 barrels of water for each barrel of oil.[10] Large projected increases in mining projects could significantly strain freshwater resources. The Canadian and Alberta governments are currently attempting to develop new and more robust rules for water use, though there is much attendant controversy; in any case, water-related constraints on oil sands producers will ultimately be determined by political authorities rather than physical limits. Resolving this issue in a way that creates a stable long-term framework, even at the expense of slower near-term expansion, will be essential to sustained growth.

Broader public concerns could also force new limits on oil sands growth. The oil sands' environmental impacts extend beyond climate and water: mining developments, in particular, require substantial forest clearing and generate large "tailings ponds" in which toxic wastes from the oil sands operations are stored. Social dislocations from booming oil communities have attracted concern; so has the upward pressure on the Canadian dollar from the oil sands boom, which (until oil dropped and the dollar weakened) hurt export-oriented industries in the rest of the country.[11] But political support for oil sands growth appears to be broad, with the center-left Liberal Party of Canada, historically seen as unsupportive of western Canadian interests, promoting the oil sands almost as vocally as the right-of-center (and more western-based) Conservative Party of Canada and the invariably pro-oil and ardently free-market Alberta government. Attitudes may change over time, but a fundamental shift appears to be unlikely.

IMPACT ON WORLD OIL MARKETS

Oil sands exploitation will affect world oil markets by displacing production elsewhere or by moderating oil prices or, most likely, through a combination of the two. New regulatory barriers to oil sands production, meanwhile, may increase world oil prices or shift production elsewhere, including to OPEC countries, though both would happen only under limited circumstances.

Analysts' predictions of the impacts' details differ depending on how they expect OPEC to behave. If OPEC (or some subset of OPEC countries, most importantly Saudi Arabia) can steer world oil prices by shaping their own production, then OPEC decisions (and those of its constituent states) fundamentally determine oil prices. In this model, increased oil sands production forces OPEC countries either to produce less at a given price (in order to maintain the same total volume of oil on the world market) or to accept a lower market price for a given amount of production (raising demand and thus accommodating more supply); this is true over both the short and long terms.[12] If, in contrast, OPEC no longer has enough cheap oil and internal cohesion to set long-term prices, then greater oil sands production simply lowers longer-term world prices by increasing the supply available at any price. (If increased oil sands production lowers world oil prices, it will also decrease production outside OPEC; the magnitude of that effect will depend on prevailing oil prices and other conditions.)

The cost of oil sands production can also play a special role in setting world oil prices. At long-term oil prices within the span over which most new oil sands projects are expected to become profitable—about $60 to $70/bbl—there are few new opportunities to produce oil.* Since oil sands provide the marginal barrel in this price range, producers can pass on part of any new costs (including environmental compliance costs) to consumers through higher world oil prices. Some relatively high-cost producers, though, may not be able to pass enough of the new costs along to be profitable, resulting in forgone investment and lower production.

At prices above about $80 per barrel, the effects of additional environmental compliance costs on production volumes and on world prices will probably be minimal. Costs for inputs like labor and oil supply services are likely to escalate, making marginal costs to producers still equal to the oil price. In this price range, though, additional burdens are much more likely to be absorbed by producers and others on the supply side (such as workers) than they are to be passed on through prices or to result in lower production.

In this context, it is important to note that world oil markets are far from perfect, particularly for unconventional oil. The cost of selling oil sands products in the United States and Canada will generally be lower than that of selling them elsewhere. The United States is the closest market for the oil sands, which keeps transportation costs down; in contrast, shipping oil sands products to Asia would require new, technically challenging, and expensive pipelines to the Pacific coast to be built. The United States also has a large amount of refining capacity—particularly

---

*Those new opportunities exist mainly in oil sands, enhanced oil recovery (EOR), and ultradeepwater production. This range coincides with recent OPEC price targets, though the credibility of those targets is questionable.

for heavy oil—that allows oil sands crude to be processed with limited or no additional investment; the same is not true for other potential markets such as China. If U.S. or Canadian policy forces oil sands crude to be sold elsewhere, and if the costs associated with shifting markets can be passed on to consumers, they will be reflected in higher world oil prices.

Energy Security and Climate Change

The prospect of sourcing oil from a stable, friendly, nearby country is naturally appealing to U.S. policymakers. It is particularly attractive to those who promote an "energy independence" agenda, which focuses on strengthening U.S. security by eliminating imports from hostile and unstable states. The purported benefits of Canadian oil are viewed more skeptically by those who emphasize that oil is traded on global markets; they place less importance on where oil is produced and more on some mix (which varies by analyst) of broadly expanding access to oil and alternatives and cutting U.S. consumption. Understanding the actual security benefits of increased oil sands production is essential to developing policy that balances those with the related climate damages.

This report does not choose a single measure of energy security. Instead, it identifies six often-articulated potential negative security and economic consequences of oil consumption and production for the United States, and assesses the impact of oil sands growth in each dimension:[13]

1. Oil revenues empower exporting states whose interests often conflict with U.S. interests.
2. U.S. economic growth is hurt by oil price volatility.
3. U.S. economic growth is hurt by wealth transfers to some oil producing states.
4. Barriers to well-functioning oil markets, including but not restricted to price manipulation by OPEC or by national governments, raise oil prices and hence hurt the U.S. economy.
5. The United States is potentially vulnerable to supply disruptions resulting from states' decisions to withhold oil supplies from world markets or from damage to oil supply chains by nonstate actors or natural disasters.

6. Dependence on oil from unstable regions may necessitate military expenditures to ameliorate risk.

The main energy security analysis is set against a business-as-usual backdrop; it is followed by a brief look at how the picture would change in the context of policy efforts or technological changes that greatly decreased U.S. or global oil demand. The report also examines how the oil sands' demands on natural gas supplies might affect the U.S. economy and U.S. security.

This analysis of energy security is followed by an assessment of oil sands' climate change consequences. That leads to principles for balancing the two goals.

OIL REVENUES EMPOWER ADVERSARIAL STATES

Revenues from oil sales can empower adversaries in two ways. They can finance spending on hostile activities. More subtly but perhaps more dangerously, they can also lessen the value to states of participating responsibly in the international economic system, blunting the tools of economic statecraft on which the United States and its allies often depend. Iran, for example, was able over the last half-decade to finance its nuclear program, weather international sanctions, and ignore incentives such as a pathway into the World Trade Organization (WTO) in substantial part because of its oil revenues.

If, over the long term, Canadian oil sands growth displaces production in places like Iran or Venezuela, or drives down the prices that such states receive for each barrel of oil they sell, it will weaken them. That is true regardless of whether the Canadian oil is consumed in the United States, since oil is to a reasonable approximation priced and sold on a global market.

Indeed, the analysis above of how oil sands production affects world oil markets indicates that increased oil sands production will lower aggregate OPEC revenues. Precise predictions are difficult, but the scales involved are easy to understand. If, for example, increased oil sands production displaces 2 mb/d of OPEC production at $100/bbl of oil, that would lead to $70 billion in reduced annual wealth transfers to OPEC producers. If that same 2 mb/d of production did not affect OPEC production but lowered oil prices by $4/bbl from a baseline of $100/bbl, it would have cut OPEC revenues by about the same amount.[14]

The extent to which any loss is borne by states like Iran or Venezuela rather than by others such as Saudi Arabia depends on how OPEC is able to function. If OPEC (or its member states) respond to increased Canadian production by cutting volume, more disciplined producers like Saudi Arabia lose most of the revenues; if it lets prices fall, producers like Iran will suffer more. In either case, the effect is modest as a fraction of these states' total oil revenues.

PRICE VOLATILITY

Increased oil sands production would do little to address short-term oil price volatility and its economic impact. There are two sides to this issue. Since oil is traded on a global market, the effects of volatility are reflected in the price of every barrel of oil regardless of its origin. This problem can be addressed only by making the U.S. economy more resilient to oil price swings, which includes—most significantly—lowering total U.S. oil consumption.

Oil sands exploitation will have a greater impact in decreasing oil price volatility in the first place, though that effect will still be limited. The oil sands, even after robust expansion, would comprise a relatively small fraction of global oil supply (perhaps 5 percent), and hence would probably have limited impact on short-term oil price dynamics. In addition, since oil sands projects are capital intensive, the oil sands do not lend themselves to building up the sort of slack capacity that can be used to smooth oil supply and hence prices.[15] That said, to the extent that oil sands crude substitutes for oil from less stable parts of the world in meeting demand, world oil prices will be less exposed to volatility arising from those less stable sources.

WEALTH TRANSFERS

Importing oil from Canada rather than from the Middle East would have another important economic benefit: it would likely decrease the U.S. current account deficit (though not by as much as simply cutting U.S. consumption). Money spent on Middle Eastern oil that is then used to purchase goods is unlikely to be spent on goods in the United States.[16] In contrast, a greater fraction of money used to buy Canadian oil will likely later be spent directly on U.S. goods and services and hence contribute directly to U.S. growth. (Money used by Middle Eastern consumers to purchase goods outside the United States can also boost the U.S. economy through trade, but the benefit will likely be less than that from direct Canadian consumption.) When such petrodollars do return directly to the United States, it has often been in the form of asset purchases, which is at best a benign phenomenon, but may carry negative national security ramifications.[17] Money can also return directly through purchases of U.S. debt; some have argued that such dynamics were an important contributor to the recently ended U.S. financial and mortgage bubble.[18]

MARKET BARRIERS

Expanded access to oil-rich areas anywhere in the world helps moderate oil prices, both by simply expanding supply and by providing diversity that serves as a hedge against disruptions. As a result, greater access is generally good for the U.S. economy. Restricting production from the Canadian oil sands, including through climate policy, would impose costs on U.S. consumers because higher oil prices would be needed to stimulate a mix of greater conservation and higher production elsewhere. So long as any restrictions were not sudden, however, this effect would likely be limited, though it would still be real. At the same time, if restrictions were seen as arbitrary or unjustified, they would undermine the broader U.S. goal of promoting open oil markets worldwide, with deeper implications for oil availability and price.

Strong growth in the oil sands would also diminish the market power of both OPEC and of individual governments that control large amounts of oil, though again the effect would be limited. If a large market player inflates prices, it will affect the price of every barrel of oil, not just those that a particular source produces; shifting to Canadian oil would not change that. But to the extent that Canadian production eroded the market share of OPEC or of individual countries, it would tend to weaken them. The effect would be both direct (smaller market share translates into less market power) and, in the case of OPEC, indirect, because with less oil production to divide up amongst its members, internal divisions would likely increase.

VULNERABILITY TO MAJOR DISRUPTIONS

Concern about supply disruptions have traditionally focused on the possibility of deliberate cutoffs by producing governments. In this respect, Canadian oil is clearly superior to oil from adversarial countries.[19] But the odds of a hostile government suddenly and deliberately cutting off supplies is small too. With world oil markets able to fairly efficiently reallocate supplies, and strategic petroleum reserves able to buffer short-term disruptions, the oil weapon is far less powerful than it once was; as a result, states are far less likely to use it. Thus, in this dimension, importing oil from Canada offers little advantage.

Canadian sources are, however, more secure than many alternatives against supply chain disruptions from nonstate actors and terrorists in particular. Such disruptions can be far more damaging than decisions by states to withhold supplies: depending on the level of damage, it can take a long time to restore elements of the supply chain. While critical infrastructure in the United States and Canada is by no means invulnerable, it is generally believed to be more secure from nonstate actors than analogous infrastructure in the Middle East or in other unstable places, such as Nigeria. Supply chains based exclusively in Canada and the United States are superior from this particular security standpoint. Unless they are extremely large, though, supply disruptions will manifest themselves primarily in global price hikes rather than physical shortages because the United States will seek oil elsewhere; still, they are not unimportant.

Canadian sources provide little if any protection against vulnerability to weather. One might imagine otherwise: events such as hurricanes Katrina and Ike have repeatedly disrupted refining in the Gulf of Mexico, and while some Canadian oil sands products will be piped to the Gulf for refining, it will be more natural over time for their refining to be concentrated in the northern United States or in Canada, areas that are less vulnerable to extreme weather.[20] But U.S. and Canadian refining capacity is essentially fixed (even though individual refineries will be modified), which means that if more Canadian oil is processed in less weather-exposed areas, more oil from other sources will be refined elsewhere; the overall vulnerabilities will remain.

MILITARY EXPENDITURES

Many have argued that U.S. military expenditures are higher than they would otherwise need to be if the United States did not depend on a relatively stable Middle East to control oil prices. They thus argue that shifting to supplies from other parts of the world would allow the United States to cut its military budget and draw down its defense commitments. If this logic were true, increased production from the Canadian oil sands would help such a shift (though the effect would be limited—Middle Eastern oil production is much larger than Canadian production will ever be). But the underlying argument is weak. While U.S. commitments in the Middle East may have strong historical ties to oil, current U.S. commitments are anchored in other fundamental problems. In particular, the long-term challenges posed by transnational terrorism, by Iran's pursuit of nuclear weapons, and by threats to Israel's security will require strong U.S. security commitments in the Middle East regardless of whether oil is also a major regional concern.

A LOW-OIL-DEMAND WORLD?

The analysis above is set against a business-as-usual backdrop in which global oil demand rises over the next two decades and U.S. consumption remains roughly flat. Public policy and new technologies might, however, lead to lower oil demand. Oil sands would be less important in a low-demand world, but unless cuts in demand were genuinely global—far from a guaranteed outcome—significant security value in a robust oil sands industry would remain.

Imagine that the United States sharply cuts its oil consumption, but other countries, particularly in the developing world, continued on their business-as-usual trajectories. The negative impacts of U.S. oil consumption on the U.S. economy—including those that result from exposure to price volatility and to market manipulation—would be reduced by lower consumption. (This is a major reason why reducing U.S. oil consumption is important.) At the same time, such reduced consumption would be unlikely to alter oil prices enough to have a large long-term impact on oil sands production. However, since the macroeconomic problems posed by oil consumption would be reduced, any value of oil sands production in mitigating those problems would be smaller too. Robust oil sands production would still have real value, though, in reducing financial flows to adversaries, since it would still displace other barrels from world markets and reduce global oil prices.

Now imagine a scenario in which large cuts in oil consumption are seen not only in the United States but also worldwide. (This is only plausible over multidecade timelines, and would still be difficult to achieve.) The results are similar to those in the previous scenario except that global oil prices would be expected to see much steeper declines. (The price here is that received by producers, not the one paid by consumers, which may be considerably higher if demand is suppressed through fuel taxes.) Production from the Canadian oil sands could be significantly diminished (relative to business as usual) if oil prices are held to low levels. Under such circumstances, however, revenues to adversarial oil producing governments would already be greatly reduced, even if their production volumes were unchanged. In this case the oil sands would, over time, become much less important for all dimensions of energy security.

NATURAL GAS

Oil dependence is not the only energy security issue associated with the oil sands. Producing a barrel of synthetic crude oil requires roughly 750 to 1,500 cubic feet of purchased natural gas, an amount whose energy content is equivalent to between one-eighth and one-quarter of a barrel of oil. If the oil sands impose overly large demands on natural gas, the United States and Canada would need to import more from abroad. The United States currently sources almost all of its gas domestically and from Canada, but the world's largest natural gas reserves are in Russia, Iran, and Qatar. The picture for U.S. natural gas has changed dramatically in the last year, though, with domestic supply projections at moderate prices much higher than they were just a short time ago. Thus, while the potential problem introduced by natural gas consumption is not negligible, it is also unlikely to be large. There are also steps that can be taken to ameliorate any problems.

Expectations for U.S. gas supplies have risen dramatically in the last year due to new optimism about unconventional gas resources. The EIA (whose projections are conservative) projects that domestic production will remain stable through 2016 and then increase steadily to 23.7 trillion cubic feet (annually) by 2030, 4.2 trillion cubic feet higher than its 2008 projection. While such projections are highly speculative, this *change* is several times the expected increase in gas demand from the oil sands (discussed in the previous section of this report), which suggests that demands from the oil sands are unlikely to be a dominant force in North American natural gas markets. Oil sands producers also continue to improve their operations' energy efficiencies and to explore alternatives to natural gas for parts of their operations. Several of these are discussed in greater detail later, but all have some prospect of decreasing natural gas demands (though with varying greenhouse gas implications).

The security consequences of U.S. dependence on natural gas are also more limited than those of dependence on oil. Oil is a problem in large part because its use is heavily concentrated in transportation, where there are few substitutes available; in contrast, a wide variety of alternatives to natural gas exist. Natural gas is also not manipulated by a cartel in the same way that oil is (though it is often manipulated by individual countries). It is also important to not extrapolate too quickly from European problems with Russia: Europe depends on pipelines, while any U.S. shift to suppliers from outside North America will depend on more flexible liquefied natural gas (LNG), which introduces fewer security problems.

OVERALL ENERGY SECURITY ASSESSMENT

The energy security benefits of robust Canadian oil sands production are real, but, because oil is essentially traded on a global market, not as large as some might intuitively assume. Oil sands exploitation will not fundamentally change the global oil picture. Perhaps the greatest impact of expanded oil sands exploitation would be a diversion of revenues away from adversarial governments—an important outcome—though this benefit would exist regardless of whether the United States was the ultimate consumer. In addition, the United States would benefit from buying oil from a country that would spend more of the proceeds on U.S. goods, and world oil markets would also gain from shifting to supply chains that are less vulnerable to terrorism. That said, U.S. vulnerability to oil price volatility and to price manipulation by OPEC or any large individual producer will not be significantly diminished by shifting imports to the oil sands, nor will the need for U.S. military commitments in the Middle East decline. The security value of oil sands production would be reduced in a world where U.S. oil consumption was cut sharply, but its role in lowering revenues to oil producing states would remain. If global consumption and prices were strongly cut over several decades—a desirable but difficult-to-promote outcome—the ultimate role of oil sands in promoting energy security would be much smaller.

CLIMATE CHANGE CONSEQUENCES

Oil sands' life cycle greenhouse gas emissions—the emissions entailed in production (including upgrading if applicable), transport, refining, and ultimate use—are greater than those associated with conventional oil. Table 1 compares the emissions from oil sands with those from other major sources of U.S. oil. The average life cycle emissions associated with a barrel of oil sands crude currently exceed those from the average barrel of oil consumed in the United States by about 17 percent.* This is due mainly to emissions from production and upgrading, which are nearly three times higher for the average barrel of oil sands crude than for the average barrel of oil consumed in the United States.[21] Actual emissions from individual oil sands projects vary widely: according to a recent RAND study, oil sands' production and upgrading emissions range from 70 kg to 130 kg per barrel; this is equivalent to exceeding the life cycle emissions from the average barrel of oil consumed in the United States by 50 kg to 110 kg per barrel, or 10 percent to 20 percent.[22] Other sources from a diverse range of viewpoints provide similar estimates.[23] Average oil sands production emissions could increase with a

---

*Some contend that this number should be lower, arguing that the appropriate reference point is the marginal alternative barrel on the world market, which is heavier and more sour than the average barrel consumed in the United States. Others would argue that oil conservation or new low-carbon fuels are the appropriate alternative, which make oil sands' relative emissions much higher.

TABLE 1. AVERAGE PER-BARREL EMISSIONS RELATIVE TO THE AVERAGE BARREL CONSUMED IN THE UNITED STATES

Source	Production, Upgrading, and Transport to Refinery	Refining and Finished Fuel Transport	Total Well-to-Tank	Total Well-to-Wheels
Canada (Oil Sands)	252%	135%	185%	117%
Venezuela (Bitumen)	221%	129%	168%	114%
Nigeria	300%	57%	162%	113%
Mexico	96%	159%	131%	106%
Angola	202%	69%	125%	105%
Kuwait	70%	135%	107%	101%
Iraq	76%	122%	102%	100%
Venezuela (Conventional)	66%	129%	101%	100%
Canada (Conventional)	88%	107%	98%	99%
Ecuador	89%	103%	97%	99%
Saudi Arabia	63%	119%	95%	99%
Domestic	62%	82%	73%	94%
Algeria	95%	46%	68%	94%

Note: Figures are based on the average for U.S. imports from each source. Emissions in the first column normally occur in the country where crude is produced; emissions in the second column are much more likely to occur in the United States. Higher figures indicate relatively dirtier sources.

Source: Based on numbers for diesel in Kristen J. Gerdes and Timothy J. Skone, *Consideration of Crude Oil Source in Evaluating Transportation Fuel GHG Emissions* (Washington, DC: National Energy Technology Laboratory, 2009).

shift from natural gas to dirtier process fuels like coal or raw bitumen, or decrease due to technological improvements; the latter trend has recently dominated.

The roughly 1.2 mb/d of current oil sands production is thus responsible for a premium of about 40 million tons of CO_2 emissions each year compared to conventional oil.[24] This is equal to about 5 percent of Canadian emissions, 0.5 percent of U.S. emissions from energy use, and slightly less than 0.1 percent of global emissions—a small piece of the emissions picture. If oil sands production increases as expected and the emissions entailed in producing each barrel are not reduced, that contribution will roughly triple by 2030, making oil sands a huge relative contributor to Canadian emissions but still a relatively marginal one in the U.S. and global contexts. If, however, policy efforts manage to slash other emissions, as they must if ambitious goals for reducing the risk of catastrophic climate change are to be met, the relative prominence of the oil sands would greatly increase. Imagine, for example, that oil sands emissions rose as expected over the next two decades and then stabilized in 2030, while total U.S. and Canadian emissions dropped by 80 percent by 2050 (an oft-proposed target). Oil sands emissions would then become equivalent to about 10 percent of U.S. emissions by 2050, representing almost all emissions from Canada at that point. Oil sands' emissions will thus be critical to deal with in the long term though not as important in the immediate future.

To get a sense how emissions controls might affect the future of the oil sands, imagine the effects of several carbon prices on the cost of producing a barrel of synthetic crude oil and on the cost of ultimately producing fuel from the oil sands. Table 2 compares the potential cost to producers for various U.S. and Canadian sources. (Other sources of oil consumed in the United States are unlikely to face carbon costs at the production stage in the near term.)

Carbon prices of $20/tCO$_2$e are similar to those that European firms currently face; most expect to see U.S. carbon prices in this range

TABLE 2. AVERAGE CARBON COST TO PRODUCERS
(ASSUMING NO ABATEMENT MEASURES)

Source	Carbon Price ($/tCO$_2$e)		
	20	50	100
Canada (Oil Sands)	$2.21	$5.53	$11.07
Canada (Conventional)	$0.52	$1.29	$2.58
United States	$0.36	$0.90	$1.80

Note: Cost will vary considerably among projects. Actual compliance costs may be lower.

Source: Underlying emissions figures are based on Gerdes and Skone, *Consideration of Crude Oil Source*.

in the near term if the United States imposes a cap-and-trade system or carbon tax, though actual prices are impossible to predict. Prices should rise toward $50/tCO$_2$e in the 2020 to 2030 time frame and continue to rise in later decades. The expected carbon costs for oil sands projects are small relative to the expected price of a barrel of oil; this is in stark contrast, for example, with coal-fired power, whose cost would increase sharply even for modest carbon prices. Still, carbon costs could affect production and pricing at the margin, and very high carbon prices in the near term could have much larger impacts.

Some argue that the burden of a carbon price on oil sands would be greater than what is suggested in Table 2, since oil sands crude will also face carbon costs at the refining stage. Table 3 shows the carbon

TABLE 3. AVERAGE CARBON COST TO PRODUCERS AND REFINERS

Source	Carbon Price ($/tCO$_2$e)		
	20	50	100
Canada (Oil Sands)	$3.96	$9.90	$19.81
Venezuela (Bitumen)	$1.55	$3.87	$7.75
Nigeria	$0.69	$1.72	$3.44
Mexico	$1.92	$4.81	$9.61
Angola	$0.83	$2.07	$4.14
Kuwait	$1.63	$4.08	$8.16
Iraq	$1.47	$3.67	$7.34
Venezuela (Conventional)	$1.55	$3.87	$7.75
Canada (Conventional)	$1.95	$4.89	$9.77
Ecuador	$1.25	$3.12	$6.23
Saudi Arabia	$1.44	$3.61	$7.22
Domestic	$1.57	$3.93	$7.86
Algeria	$0.56	$1.40	$2.80

Note: This assumes that no abatement measures are taken. Carbon pricing assumed to apply only inside the United States and Canada.

Source: Underlying emissions figures are based on Gerdes and Skone, *Consideration of Crude Oil Source.* The product is assumed to be diesel; the figures would change little for gasoline.

cost for a wide range of sources if prices through refining are considered. It assumes that Canadian and U.S. producers face carbon costs throughout and that others face carbon costs for refining and final distribution in the United States. (The possibility of more refining shifting to unregulated markets is real but is beyond the scope of this study.) The cost to Canadian sources roughly doubles but the cost to other sources—including direct near-term competitors like Venezuelan bitumen—increases, often substantially, too. Whether this extra cost will be absorbed by producers (through reduced production or lower profits), by refiners, or by consumers (through higher prices) depends on the finer details of refining and product markets and is extremely difficult to predict. The possibility that the impact of carbon pricing will be larger than what is indicated in Table 2 is, nonetheless, impossible to dismiss.

BALANCING ENERGY SECURITY AND CLIMATE CHANGE

The preceding sections make clear that oil sands production delivers energy security benefits and climate change damages, but that both are limited. A healthy balance is possible. Global economic conditions along with Canadian policy will be the main determinants of the oil sands' future, but U.S. policy will play a critical role.

For the near future, the economic and security value of oil sands expansion will likely outweigh the climate damages that the oil sands create—but climate concerns cannot and must not be ignored, and will become more important over time. U.S. policymakers should balance the two goals by working with Canada to promote strong incentives to cut the emissions associated with each barrel produced from the oil sands, without directly discouraging production itself. They should also seek to avoid measures that would promote increases in global (pretax) oil prices unless such measures deliver clear countervailing climate benefits. Since the oil sands are a limited piece of the energy and climate puzzles, any policies will need to be embedded in a much broader strategy to cut global emissions and to increase U.S. energy security.

Policy Recommendations

The final section of this report makes recommendations for U.S. policymakers, emphasizing opportunities for the United States and Canada to work together. It begins by outlining the current state of policy in both countries. It then makes recommendations on carbon markets, emissions standards for transportation fuels, technology policy, and other regulatory tools.

CURRENT POLICY

Canadian and Albertan policy currently play a much larger role in shaping the oil sands' development than U.S. policy does. Alberta has relatively low royalty and tax rates for the oil sands, which promotes greater production. The Alberta government has also imposed a carbon tax of $15/tCO$_2$e on oil sands producers (among others), which applies to emissions above 88 percent of those producers' historical per-barrel average.[25] That price is fairly low, but it is also higher than any explicit carbon price faced by any industry in the United States; it also discourages producers from switching from natural gas to higher-emissions fuels. Revenues go into a technology fund that focuses on, among other things, developing carbon capture and sequestration (discussed in more detail below).

Over the longer term, Canadian and Alberta government policy is likely to impose steadily tougher but still supportive rules for oil sands development and for climate change in general. Federal and provincial policy (including the federal opposition party stance) is broadly supportive of the oil sands as boosting Canadian economic growth and international influence. At the same time, while Canada will not meet its Kyoto targets, the federal government has proposed cutting national emissions by 20 percent from 2006 levels by 2020, and the main opposition party has called for more aggressive goals without being specific about its preferences. (Unlike the United States, Canada ratified the Kyoto protocol.) Both the Canadian government and the main opposition party support integration of U.S. and Canadian cap-and-trade systems and oppose other U.S. regulations that would interfere with oil sands development.

Perhaps the greatest political uncertainty surrounds other elements of environmental policy: the current opposition party is more likely to support regulations on water and other impacts that could constrain oil sands development (or add new costs) than the current federal or Alberta governments would support. Regulation of local environmental impacts, though, is most likely to be driven at the provincial level, where the Progressive Conservative Party of Alberta, skeptical of environmental regulation and in power since 1971, is likely to remain in control.

The fact that the United States is the natural market for oil sands products means that U.S. government policy will inevitably affect the oil sands. The United States will also affect the context for oil sands development through its interactions with Canada in international climate negotiations. U.S. government policy currently has little influence on the oil sands in either dimension. The greatest current exercise of U.S. power as a consumer is through a provision in the Energy Independence and Security Act of 2007, which bars U.S. agencies from entering into contracts to procure unconventional oil that entails higher life cycle CO_2 emissions than conventional petroleum. This provision was aimed at stopping the U.S. Air Force from contracting for production of liquid fuels made from coal, but it technically restricts purchase of fuels derived from oil sands as well, and as a result has attracted much attention on both sides of the border. The legal situation is ambiguous, though, since fuel from oil sands is not procured directly by the U.S. government; it is mixed with conventional fuel (as well as other liquid fuels) and purchased on commercial markets. Some oil companies fear that refineries may be forced to avoid oil sands crude in order to sell fuel to the U.S. government, distorting markets, but that outcome is unlikely. Meanwhile, the United States does little to influence Canadian decisions through international climate negotiations. The Obama administration has focused mainly on its domestic efforts and on pressing developing countries to cut their emissions; it has not devoted significant attention to other developed countries.

Regulations currently proceeding at the state level, laws under consideration in Congress, and possible U.S. environmental lawsuits may have much greater impact. California recently decided to adopt a low-carbon fuel standard, which would require the life cycle emissions of fuel sold in the state to decrease over time; other states, particularly in the Midwest and Northeast, have signaled their intent to follow. Depending on its details, a low-carbon fuel standard could make oil sands less attractive—potentially considerably so. (This is addressed in detail below.) There is also substantial interest in Congress in some variation on a low-carbon fuel standard.

Moves in Congress to adopt a cap-and-trade system may also be accompanied by steps that would affect the oil sands. A U.S. cap-and-trade bill is likely to include measures that would eventually impose border tariffs on emissions-intensive trade-exposed goods being imported from countries assessed to have substantially weaker climate regulations; depending on the scope of that regulation, and if Canada was judged to have weak emissions standards, oil sands imports could face tariffs.* Congress is also likely to appropriate significant sums for clean technology development, some of which could be relevant to the oil sands. Some in the United States may also seek to use U.S. power as a consumer to influence Canadian policy through the permitting process for pipelines and refineries designed to handle oil sands crude.

CARBON PRICING, EMISSIONS LIMITS, AND CARBON MARKET INTEGRATION

The central tool for balancing energy security and climate change concerns in the oil sands context should be reasonable and prudent carbon pricing. That would provide polluters incentive to cut their emissions while maintaining support for open energy markets; done right, it would also avoid driving up global (pretax) oil prices or inflating the market share of low-cost producers (notably in those in OPEC). In practice, political trends suggest that both the United States and Canada are likely to eventually adopt economy-wide cap-and-trade systems. Integrating those systems is the best way to ensure that the oil sands face carbon prices that are neither too low nor too high.

---

*Border tariffs incorporated in major pieces of proposed legislation to date cover only manufactured goods and hence would not affect oil sands imports.

Policy Recommendations

There is a compelling case, even absent the oil sands, for harmonizing U.S. and Canadian carbon pricing schemes; the oil sands factor, in both its energy security and climate change dimensions, only makes that case stronger. The U.S. and Canadian economies are highly integrated. Free trade in goods and services, including energy-intensive products, is extensive. If the United States and Canada impose sharply different prices on greenhouse gas emissions, the markets in tradeable energy-intensive goods, such as steel, cement, oil, and electricity, would become significantly distorted, with production migrating to whichever country imposed more lax rules. That reason alone should motivate efforts to harmonize emissions prices.[26]

The simplest way to harmonize the prices of emissions permits in two cap-and-trade systems is to allow trading between them; this will naturally lead prices to be the same on both sides of the border.[27] This makes particular sense in emissions-intensive businesses like oil production where all the other major factors of production are already freely traded. It is vastly preferable to the confrontational approach of using border tariffs to equalize prices.

Since Canadian emissions are less than 15 percent of U.S. emissions, supply of and demand for permits in the U.S. system will be the primary determinant of permit prices in an integrated scheme. A joint system is thus likely to have permit prices similar to those expected for a U.S.-only system. The Canadian government has expressed a strong preference for a harmonized system, suggesting that a desire for predictable pricing trumps sovereignty concerns.

Indeed, carbon autarky could lead to problems since carbon prices in a Canada-only system would be very sensitive to the total cap on Canadian emissions, potentially leading to excessively high or low prices as well as substantial volatility. Imagine, for example, that Canada was pressed to cut its emissions to 1990 levels by 2020, the same goal currently being contemplated for the United States in a range of U.S. legislation. Because Canadian emissions are expected to steadily increase due to expanded oil sands operations, this is a much sharper cut relative to business as usual for Canada than for the United States.[28] As a result, carbon prices could be substantially higher in Canada than in the United States. Differing assumptions lead various models to predict different prices, but the general conclusion is robust. For example, a 2007 study by M. K. Jaccard and Associates for the David Suzuki Foundation, an environmental nongovernmental organization, concluded that driving Canadian emissions back to 1990 levels by 2020 would require a carbon price that started at $40 (in Canadian dollars) in 2010, rising to $65 in 2015 and $100 in 2020, considerably higher than most of the carbon prices anticipated in a U.S. system.[29] If the prices in a U.S. system are reasonable, then considerably higher prices in Canada are probably not—and could unreasonably affect the oil sands.

The prospect of high and unpredictable prices could, of course, discourage Canada from formally committing to emissions limits in the first place or to meeting its commitments. That strengthens the case for market integration: by linking its market to Canada's, the United States would bind Canada to controlling its emissions (and Canada could also bind the United States).

Uneven pricing could also introduce a real danger of so-called emissions leakage. Since more than half of the emissions from oil sands production come from upgrading, overly high Canadian prices would put increasing pressure on oil sands producers to ship bitumen to Asia for upgrading in unregulated markets. This move could ultimately drive up greenhouse gas emissions and would deprive the United States of some of the energy security benefits of importing oil from Canada rather than from other parts of the world. It would also increase the total costs of oil sands production, which might translate to higher world oil prices and hence to greater windfalls for low-cost producers. Canadian prime minister Stephen Harper has promised not to allow exports of such products to markets without comparable greenhouse gas regulations, but that promise is far from guaranteed to hold if the United States sharply curtails its consumption.

Integrating U.S. and Canadian carbon markets would entail significant challenges. Technical barriers would, as in the European trading system, be minimal. Integration would, however, require harmonization of a range of trading-system features, including rules for carbon offsets as well as any price floors and ceilings. Most critical, any aggregate emissions cap would need to be divided into initial national targets so that each government could either distribute or auction the emissions permits. The choice of division would not affect the price of carbon, but it would have major ramifications for each country's revenues as well as the ability of each government to compensate hard-hit domestic players by allocating them free permits.

The U.S. approach to allocating emissions permits should, in the near term, ensure that Canada has enough permits to freely allocate a substantial number to oil sands producers (proportionate to their oil output, not their emissions, in order to maintain a strong incentive for them to cut those emissions). Indeed, the United States should encourage such a Canadian approach. If short-term prices are in the $30 to $40 range or if anticipated long-term oil prices are in the range of roughly $60 to $70 per barrel, imposing new costs on oil sands production will either increase per-barrel windfall profits to low-cost producers, most notably in OPEC, or reduce oil sands production and hence, allow those low-cost producers to sell more oil, again reaping greater windfalls. There is a risk, of course, of delivering a windfall to oil sands producers, but it is outweighed by the reduced risk of doing the same for adversarial oil producers.[30]

Oil prices are ultimately likely to be far higher than those necessary to stimulate investment in the oil sands; at that point, oil companies will be able to absorb the cost of emissions permits, and any free permits will be unnecessary. (At, for example, an oil price of $100/bbl and a high carbon price of $100/tCO$_2$e, oil sands producers would face extra costs of between $10/bbl and $20/bbl, which they would probably not be able to pass on to consumers but would likely be able to absorb without curtailing production.) A scheme that scaled back free allowances at higher oil prices would mirror the approach taken in Alberta's recently revised royalty policy, which ramps up royalty rates as oil prices increase. If long-term prices are far lower, it will be because of deep cuts in consumption; as argued earlier, the oil sands will be far less important in such a scenario. If Canada still chooses, at that point, to give scarce permits for free to oil sands producers, that should not be a problem for the United States, so long as Canada's overall emissions targets maintain their integrity.

LOW-CARBON FUEL STANDARDS

Many U.S. policymakers, including both 2008 presidential candidates as well as more than a dozen governors from both parties, have advocated the adoption of a so-called low-carbon fuel standard (LCFS), which they view as a tool for pursuing energy security and climate change goals together. An LCFS would require that fuel refiners, blenders, and importers steadily reduce the life cycle greenhouse gas intensity of the fuel consumed in the cars and trucks they serve, which would incentivize emissions cuts while promoting low-carbon alternatives to oil. A simple low-carbon fuel standard, however, could also impose a heavy cost on oil sands because of their higher-than-average emissions. That would exacerbate energy security problems without delivering compensating climate benefits. Any standard should be designed to not discriminate between conventional and unconventional oil. This would provide the benefits of an LCFS in promoting biofuels and electric vehicles without creating energy security downsides.

Simple economics suggests that low-carbon fuel standards are inefficient tools for reducing greenhouse gas emissions. (Their energy security value is discussed below.) An economy-wide cap-and-trade system would incentivize emissions cuts wherever they were cheapest while an LCFS would demand emissions cuts from a shift in the types of fuels used in transportation, even if it was cheaper to reduce emissions through measures elsewhere in the economy (or through better fuel economy or less driving). Restricting the set of options available for meeting a policy goal generally makes achieving that goal more expensive.[31]

A poorly designed low-carbon fuel standard could actually have perverse energy security implications rather than delivering a win-win outcome. If secure low-carbon alternatives did not rapidly materialize, other secure (but high-carbon) options would be shut out too, compounding the negative security ramifications.[32] Emissions pricing policy that seeks to prudently balance energy security and climate change would be better off looking for emissions cuts elsewhere in the economy if cuts from the carbon content of fuels can't be delivered quickly. Meanwhile, as the United States seeks to strengthen energy security by curtailing oil demand, it should do that directly (through gasoline taxes, fuel efficiency standards, or otherwise), rather than as an incidental effect of a low-carbon fuel standard that has other undesirable side effects.

The most common counterargument is that this misses the real goal of an LCFS. A low-carbon fuel standard would create a clear and sizable market for advanced biofuels and electric vehicles in a way that avoids some problems associated with traditional mandates, does not pick specific technological winners and losers, and ensures that emissions are being reduced. That would allow investors to scale up the most promising biofuels production and electric vehicle investments with confidence, which in turn would bring the associated costs (along with the cost of cutting emissions) down. A simple economy-wide price on carbon, whether delivered through a cap-and-trade system or a carbon tax, would not address this potential market failure.

Whether an LCFS is an effective way of driving down the cost of gasoline alternatives is beyond the scope of this report. But if the United States does adopt a low-carbon fuel standard, or if the U.S. government allows states to do so, that standard should not distinguish between different fossil-based sources.* A standard crafted this way would provide a win-win outcome: it would still expand biofuels and electric vehicle markets but would not penalize unconventional oil. The balance between conventional and unconventional oil is most appropriately addressed through carbon pricing as well as through other channels.

LOW-CARBON TECHNOLOGIES: CARBON CAPTURE AND SEQUESTRATION AND NUCLEAR POWER

Carbon capture and sequestration (CCS) and nuclear power have both been touted as potential saviors for the oil sands. As carbon prices steadily rise over the coming decades, moving to such technologies may become wise. The Canadian and Alberta governments should be encouraged to intensify their support for such alternatives, which would help meet their economic and climate objectives. U.S. funds, though, would be better spent elsewhere.

The Alberta government has devoted $2 billion to support CCS research, development, and demonstration (RD&D) projects. Several companies have applied for funding; the applications are expected to be processed in the next two months. Others are moving ahead independently: for example, Enhance Energy, an Alberta-based oil company, recently filed regulatory applications for Alberta's first CO_2 pipeline, which it claims will be operational by 2011.[33] (The CO_2 would be used for enhanced oil recovery.) A 2008 RAND study concluded that CCS could technically cut life cycle emissions from the oil sands to marginally less than life cycle emissions associated with light sweet crude. Ultrahigh carbon prices will have no special effect on the oil sands if CCS can be deployed at reasonable cost. Development of CCS would also indirectly help address any problems arising from the oil sands' demands on natural gas, since it would allow a shift from natural gas to

---

*This argument naturally extends itself to other unconventional oil, including coal-to-liquids. It should not be read, however, as an endorsement of that technology by the author, who expects that a reasonable carbon price would already (legitimately) create problems for coal-to-liquids as an alternative.

coal, raw bitumen, or waste for steam production without a significant corresponding emissions penalty.

But what are the prospects for CCS—and at what pace and price? Opinions vary widely. RAND estimates that costs are $3.71 to $7.88 per barrel for mining projects and $5.67 to $10.80 for in situ; they project that those costs will drop to $2.86 to $6.39 and $4.36 to $8.73, respectively, by 2025.[34] (Many believe that the costs will be higher.) The bulk of the costs (and potential cost improvements) come at the capture stage. With these costs, CCS would be useful if carbon prices rose to about $40 to $80 per ton of carbon dioxide, something that will need to happen over the medium to long term. Supporting development of CCS, then, would help insulate the oil sands from high carbon prices in the long term as a contributor to U.S. energy security.

Many have expressed skepticism about the value of CCS for the oil sands. Much has been made of a "secret" memo for Canadian government ministers that critics claim shows that CCS will fall far short of fixing the oil sands' carbon problems. They cite a passage stating that "only a small percentage of emitted CO_2 is 'capturable' since most emissions aren't pure enough. Only limited near-term opportunities exist in the oil sands and they largely relate to the upgrader facilities." This is technically true but it has been misinterpreted. Only "a small percentage of emitted CO_2 is pure enough to be capturable" because most emissions occur when the fuel produced from the oil sands is used in driving.[35] But these emissions are common to all gasoline—they do not present a special problem for the oil sands. Moreover, within oil sands production, upgrading is the largest source of emissions, comprising roughly 50 percent to 65 percent of total production emissions. (Other point sources—more diffuse but not necessarily unmanageable—comprise another 20 percent to 45 percent of emissions.) Eliminating emissions from upgrading would bring emissions from oil sands production much closer to emissions from production of conventional oil. Finally, the oil sands' emissions problems are more pronounced in the long term than in the short term; the fact that CCS may not fix their emissions problem soon is not enormously important.

Serious observers have also proposed using nuclear power as a substitute for natural gas in the oil sands. This has the potential in principle to alleviate both energy supply and emissions concerns. But its practicality is questionable. The large scale typical of nuclear plants matches poorly with the distributed nature of oil sands operations. While a range of options can in principle provide electricity for mining operations and both electricity and hydrogen for upgrading, the CANDU nuclear reactors typical in Canada are too large to provide steam for SAGD projects. Smaller pebble bed modular reactors appear to be properly sized for steam production.[36] The appropriate pebble bed reactors, though, are not yet commercially available.

The typical lead time for nuclear projects is also even longer than that for oil sands projects. Uncertainty in projected oil sands growth would thus introduce significant risks for nuclear developers. Despite several years with much noise about the potential for nuclear power in the oil sands, no license applications for reactors have yet been filed. Again, though, pebble bed reactors might address this issue in the longer term, since their construction times might be relatively short. That said, it would be unwise to bet the oil sands' future on nuclear power.

What does this mean for U.S. policy? Despite the value of CCS in particular for oil sands operations, significant direct U.S. financial support for oil sands CCS development is inappropriate. The Canadian and Albertan governments will invest substantial amounts in RD&D for CCS in oil sands applications because of its economic importance to them. (The United States and Canada should work together to ensure that CCS investments are sufficient, given the consequences to both the Canadian economy and U.S. energy supplies.) But political appetite in the United States for anything that can be described as a subsidy for fossil fuel industries will be limited. U.S. policymakers should focus their fossil RD&D money on CCS for power generation, where the emissions involved dwarf oil sands emissions and where cooperation with Canada (as well as others) would be valuable. Some learning between the two efforts is possible; it is, however, likely to be limited. The primary challenges in scaling up CCS are not in the individual components but in technology integration, establishing working business models, and crafting regulatory approaches. The challenges in those dimensions promise to be very different in the oil sands and power generation contexts, limiting the extent to which learning can be usefully shared between the two.

OTHER REGULATORY RESTRICTIONS

While most U.S. policy decisions relevant to the oil sands focus on broad carbon regulations (cap-and-trade or low-carbon fuel standards), other regulatory decisions are also relevant. In particular, some may seek to use the permitting process for pipelines and refineries to target infrastructure that is being built to handle oil sands products. Assuming oil sands operations face similar incentives to cut emissions to other activities, the U.S. government should aim to ensure other U.S. regulatory processes are not used to block oil sands imports on climate change concerns or grounds related to local social or environmental impacts in Canada.

The most prominent effort to date to use the U.S. regulatory approval process to block oil sands-related infrastructure is a lawsuit filed by the Natural Resources Defense Council (NRDC) against the U.S. Department of State in August 2008. (The State Department is responsible for approval of pipelines that cross international borders.) NRDC argues that the Environmental Impact Statement produced by the State Department of the last U.S. administration for TransCanada's Keystone pipeline neglects to consider the impact of refining and extracting oil sands crude on greenhouse gas emissions, even though the extraction emissions occur outside U.S. borders. The State Department has responded that only the environmental impacts of actually transporting crude through the pipeline should be considered.

While there is a legitimate argument that, in the absence of economy-wide carbon pricing, such blunt approaches to effecting change may be necessary, carbon pricing through a cap-and-trade system is vastly preferable. And while some may want to use the U.S. regulatory process to pressure Canada and Alberta to address other environmental and social impacts of oil sands operations, decisions in those dimensions are most appropriately left to those directly affected in Canada and Alberta. The State Department should continue to oppose inclusion of greenhouse gas emissions in its permitting process, and the federal government should encourage individual states to behave similarly where necessary.

Beyond this, the Energy Independence and Security Act of 2007, described earlier, should be legally clarified to ensure that it does not block the U.S. government from buying commercial fuels derived in part from oil sands. There is nothing wrong in principle with the U.S. government steering its purchasing decisions toward more climate-friendly products. But implementing a broad prohibition on U.S. government purchases could create havoc in the marketplace, requiring refineries to sharply shift their operations. (U.S. government employees, for example, buy fuel at nearly every service station in the United States.) It would also create a major irritant to the more important diplomacy required to establish uniform carbon pricing.

CONCLUSION

The steps just outlined would help ensure that energy security and climate change were effectively balanced in U.S. policies that affect the Canadian oil sands. A smart strategy should combine four basic elements:

- *Link U.S. and Canadian cap-and-trade systems.* Fair and stable carbon pricing in Canada would help both countries reap the benefits of the oil sands while mitigating their damages. Linking the cap-and-trade systems that are likely to evolve in the two countries is the best way to do that. The United States should also ensure that Canada is able to initially provide a small number of free emissions permits to oil sands producers. This would mitigate a risk of allowing carbon pricing to raise world oil prices (delivering windfalls to low-cost producers) while maintaining most incentives for oil sands operators to cut emissions.
- *Tread carefully with any low-carbon fuel standard.* The United States should design any low-carbon fuel standard—an increasingly popular potential regulation that would require specific cuts in the average emissions associated with every unit of transportation fuel—so that the oil sands, which should already face a reasonable carbon cost, are not penalized again (and perhaps much more heavily) for their higher emissions. An ill-designed scheme could burden the oil sands (along with several other U.S. oil sources) in ways that would damage U.S. energy security without providing commensurate climate benefits.
- *Focus U.S. technology support on higher-payoff areas.* There may be pressure for the United States to provide funds for research, development, and demonstration efforts in carbon capture and sequestration or nuclear power for the oil sands. While basic technical cooperation is always valuable, these would generally not be U.S. dollars well spent. The scale of the other energy and climate problems facing the United States demands that U.S. energy innovation support focus elsewhere, including in CCS for power plants (a substantially distinct problem) as well as in renewables, transmission, and efficiency.
- *Resist the misuse of other U.S. environmental regulations to constrain oil sands.* So long as the oil sands are expected to face a fair and

reasonable carbon price, the United States should resist attempts to use U.S. environmental regulations to block permitting of oil sands-related pipelines or refineries on climate grounds. Some may also try to use such regulations as a back door to dealing with the local social and environmental effects of oil sands development in Canada. Decisions on how to deal with these local effects—many of which can be disturbing—should be made by the affected communities in Canada rather than forced by outside U.S. action. The direct effects of pipelines and refineries on communities in the United States should be dealt with on the same basis as the effects of other oil-related infrastructure.

These measures must be part of a much broader strategy. The oil sands are one of many pieces of the U.S. energy security puzzle—a broader U.S. strategy must focus on cutting oil consumption (both at home and globally), ensuring access to resources along with well-functioning markets, and promoting alternatives to oil. Cleaning up the oil sands, meanwhile, is only a small part of the climate challenge, particularly in the near term. The United States will also need a far broader strategy for pursuing emissions cuts at home and around the world; as in the energy security dimension, obsession over the oil sands would be a dangerous distraction.

Endnotes

1. International Energy Agency, *World Energy Outlook 2008* (Paris: International Energy Agency, 2008), p. 215. BP estimates similar total proven reserves at 152 billion barrels as of the end of 2007. See BP, *BP Statistical Review of World Energy June 2008* (London: BP, 2008).
2. The EIA estimates that about 10 billion barrels are technically recoverable; fewer will likely be economically recoverable. *Analysis of Crude Oil Production in the Arctic National Wildlife Refuge* (Washington, DC: Energy Information Administration, 2008).
3. International Energy Agency, *World Energy Outlook 2008*, p. 262.
4. *Analysis of Crude Oil Production in the Arctic National Wildlife Refuge.*
5. David McColl, *The Eye of the Beholder: Oil Sands Calamity or Golden Opportunity?* (Calgary, AB: Canadian Energy Research Institute, 2009). The study includes a band of uncertainty that ranges from 1.9 to 3 mb/d in 2015 and from 3.7 to 5.4 mb/d in 2030. The Canadian Association of Petroleum Producers is reported to have similar projections. Claudia Cattaneo, "CERI Warns on Alberta Oil Sands Investment," *National Post*, February 5, 2009.
6. Prices in this report are in U.S. dollars unless otherwise noted.
7. Estimates in this paragraph are based on discussions with a variety of independent and industry analysts.
8. R. B. Dunbar, *Gas Use in the Canadian Oil Sands Industry* (Calgary, AB: Strategy West, 2007).
9. *Canadian Natural Gas: Review of 2007/08 & Outlook to 2020* (Ottawa, ON: Natural Resources Canada, 2008).
10. Michael Toman et al., *Unconventional Fossil-Based Fuels: Economic and Environmental Trade-Offs* (Santa Monica, CA: RAND Corporation, 2008).
11. For a sobering if sometimes overwrought (and distinctly Canadian) look at the broader social and environmental impacts of oil sands development, see Andrew Nikifourk, *Tar Sands: Dirty Oil and the Future of a Continent* (Vancouver, BC: Greystone Books, 2008).
12. Since commercial oil production is driven by economic assessments of expected costs and revenues, anticipated oil prices largely determine investment and, ultimately, production capacity. Those same oil prices also determine oil demand. Neither oil companies nor OPEC countries are very good, though, at projecting the impact of prices on supply or demand. That means that as the actual level of demand and non-OPEC supply at a given price is revealed, OPEC countries must adjust their production to maintain a target price (by bringing supply in line with demand at that price), or accept a different price from the one it originally targeted (which would allow demand to adjust).
13. This is not a complete list. Other energy security concerns include the proliferation of bilateral supply arrangements, particularly involving China, and domestic governance

problems in oil-exporting states. See John Deutch, James R. Schlesinger, and David G. Victor, et al., *National Security Consequences of U.S. Oil Dependency*, Council on Foreign Relations, 2006, Independent Task Force Report No. 58 (New York: Council on Foreign Relations Press, 2008).

14. This assumes that OPEC production is 40 mb/d, global demand is 100 mb/d, the long-term price elasticity of oil demand is (conservatively) -0.5, and other non-OPEC production is not affected.
15. States with oil that requires little up-front investment to produce can cheaply maintain standby capacity to quickly increase their production; states like Canada that have high up-front costs cannot.
16. "Ben's Bind," *The Economist*, May 1, 2008. The article cites a Goldman Sachs analyst report for this conclusion.
17. Brad W. Setser, *Sovereign Wealth and Sovereign Power*, Council Special Report No. 37 (New York: Council on Foreign Relations Press, September 2008).
18. Steven Dunaway, *Global Imbalances and the Financial Crisis*, Council Special Report No. 44 (New York: Council on Foreign Relations Press, March 2009).
19. Some might note that Canada withheld oil supplies for a period from 1973 to the mid-1980s as part of its energy security strategy. Such action is different from using the "oil weapon" and in any case is now precluded by the North American Free Trade Agreement (NAFTA).
20. To the extent that the United States decreases Venezuelan imports, more oil sands product will go to the Gulf for refining, since capacity designed to handle heavy oil will be freed up.
21. Emissions from refining are also higher than average, since oil sands crude has high sulphur content that must be removed through energy-intensive processes; this is offset by lower than average emissions from moving nearby Canadian oil to market.
22. Michael Toman et al., *Unconventional Fossil-Based Fuels*.
23. The World Wildlife Fund, for example, believes that the extra emissions from the aveage oil sands barrel, compared to a barrel of conventional oil, are about 60 kg, if not somewhat higher. World Wildlife Fund, *Unconventional Oil: Scraping the Bottom of the Barrel?* (London: WWF-UK, 2008). The Canadian Association of Petroleum Producers, meanwhile, claims that excess emissions are about 70-80 kg per barrel compared to conventional oil—a comparable figure. "Canada's Oil Sands," http://www.canada-soilsands.ca/en/what-were-doing/greenhouse-gas.aspx.
24. This is based on a conservative assumption of an average of about 0.1 tons of extra CO_2 per barrel of oil.
25. Government of Alberta, *Alberta's 2008 Climate Change Strategy* (Edmonton, AB: Government of Alberta, 2008).
26. The same argument suggests that integrating cap-and-trade systems with Mexico, something frequently discussed, could also be wise; that option is, however, beyond the scope of this study.
27. There are, in principle, other ways to harmonize prices. Canada could dynamically adjust its supply of emissions permits in order to control prices; in practice, though, this would be nightmarish to implement. Markets could also be linked indirectly, though far from perfectly, by trading with a third system.
28. This goal is far less stringent than the cuts of 25 percent to 40 percent below 1990 levels being sought by the European Union.
29. Rose Murphy et al., *Cost Curves for Greenhouse Gas Emission Reductions in Canada: The Kyoto Period and Beyond* (Vancouver, BC: MK Jaccard and Associates Inc., 2007).

30. Imagine that governments provide free allocations to producers worth $5/bbl for 2 mb/d of oil, at a cost of $10 million per day. If that avoids even a $1/bbl increase in global oil prices, U.S. consumers save $20 million per day on 20 mb/d in purchases. As the volume of oil sands production (and hence the value of free allowances) increases and as total U.S. consumption (and hence the associated savings) drops, the benefit of this strategy decreases.
31. This discussion assumes an LCFS is implemented on top of an economy-wide cap-and-trade system. The only study to estimate the cost of complying with a low-carbon fuel standard pegs the implicit carbon price at $300–$2,000/tCO$_2$e for a national LCFS that requires a 10 percent cut in fuels' greenhouse gas intensity by 2020. (Stephen P. Holland et al., "Greenhouse Gas Reductions Under Low Carbon Fuel Standards?," NBER Working Paper 13266, July 2007.) These are staggeringly large numbers, many times higher than the near-term prices forecast for the most stringent plausible cap-and-trade systems. Real costs might be much lower, but the upside risk is real.
32. An LCFS could also be very difficult to administer, requiring monitoring of emissions in the production of oil from places like Nigeria, Mexico, and Venezuela, all of which entail relatively high emissions.
33. Lynda Harrison, "Half-Billion-Dollar CO$_2$ Pipeline Proposed for 2011 Start," *Daily Oil Bulletin*, February 24, 2009.
34. These numbers do not translate directly into $/tCO$_2$e. Loosely, though, if CCS reduces emissions by about 100 kg CO$_2$e/bbl, these figures translate into a carbon cost of between $30 and $90 per barrel.
35. See CBC News, "Secret advice to politicians: oil sands emissions hard to scrub," http://www.cbc.ca/canada/story/2008/11/24/sands-trap.html.
36. Ashley Finan, "Integration of Nuclear Power with Oil Sands Extraction Projects in Canada," MIT MSc thesis, June 2007.

About the Author

Michael A. Levi is the David M. Rubenstein senior fellow for energy and the environment at the Council on Foreign Relations (CFR). He is director of the CFR program on energy security and climate change and was project director for the CFR-sponsored Independent Task Force on Climate Change. Dr. Levi was previously fellow for science and technology at CFR, and, before that, a nonresident science fellow and a science and technology fellow in foreign policy studies at the Brookings Institution. He is the author of two books, *On Nuclear Terrorism* and *The Future of Arms Control*, and is a frequent contributor to newspaper opinion pages and magazines. Dr. Levi holds a BSc from Queen's University (Kingston), an MA in physics from Princeton University, and a PhD in war studies from the University of London (King's College).

Advisory Committee for *The Canadian Oil Sands*

Jesse H. Ausubel
Rockefeller University

David P. Bailey
Exxon Mobil Corporation

Lisa B. Barry
Chevron Corporation

Robert A. Belfer
Belfer Management, LLC

Ruth Greenspan Bell
World Resources Institute

Tara Billingsley
U.S. Senate Committee on Energy and Natural Resources

Eileen B. Claussen
Pew Center on Global Climate Change

Kathleen B. Cooper
John Goodwin Tower Center for Political Studies

Richard N. Cooper
Harvard University

John Deutch
Massachusetts Institute of Technology

Rick Duke
Natural Resources Defense Council

Julian M. Flannery

Gordon D. Giffin
McKenna Long & Aldridge LLP

Peter Gottsegen
CAI Advisors

Paul L. Joskow
*Alfred P. Sloan Foundation;
Massachusetts Institute of Technology*

Ernest J. Moniz
Massachusetts Institute of Technology

Edward L. Morse
Louis Capital Markets

Steven Mufson
The Washington Post

Rodney W. Nichols

Scott Nyquist
McKinsey & Company, Inc.

Roger P. Parkinson
University of Toronto Press

Michelle Billig Patron
PIRA Energy Group

Peter Tertzakian
ARC Financial Corp.

Note: Council Special Reports reflect the judgments and recommendations of the author(s). They do not necessarily represent the views of members of the advisory committee, whose involvement in no way should be interpreted as an endorsement of the report by either themselves or the organizations with which they are affiliated.

Maurice R. Greenberg Center for Geoeconomic Studies Advisory Committee

Roger C. Altman
Evercore Partners, Inc.

Richard H. Clarida
Columbia University

Richard N. Cooper
Harvard University

Steven A. Denning
General Atlantic LLC

Daniel W. Drezner
Fletcher School of Law and Diplomacy

Nicholas Eberstadt
American Enterprise Institute for Public Policy Research

Jessica P. Einhorn
SAIS, Johns Hopkins University

Martin S. Feldstein
National Bureau of Economic Research

Joseph H. Flom
Skadden, Arps, Slate, Meagher & Flom LLP

Kristin J. Forbes
Massachusetts Institute of Technology

Jeffrey A. Frankel
Harvard University

Stephen C. Freidheim
Cyrus Capital Partners

Aaron L. Friedberg
Woodrow Wilson School of Public and International Affairs

James D. Grant
Grant's Interest Rate Observer

Maurice R. Greenberg
C. V. Starr & Co., Inc.

David D. Hale
David Hale Global Economics

Amy M. Jaffe
James A. Baker III Institute for Public Policy

Jim Kolbe
German Marshall Fund of the United States

Marc Lasry
Avenue Capital Group

Robert E. Litan
Ewing Marion Kauffman Foundation

Michael Mandelbaum
Johns Hopkins University

Donald B. Marron
Lightyear Capital

Kathleen R. McNamara
Georgetown University

Joseph S. Nye
Harvard University

E. S. O'Neal

Charles O. Prince
Sconset Group

Jeffrey A. Rosen
Lazard

John G. Ruggie
Harvard Kennedy School

Faryar Shirzad
The Goldman Sachs Group, Inc.

Maurice R. Greenberg Center for Geoeconomic Studies Advisory Committee

Joan E. Spero
Foundation Center

Fareed Zakaria
Newsweek International

Daniel B. Zwirn
D. B. Zwirn & Co., LP

Mission Statement of the Maurice R. Greenberg Center for Geoeconomic Studies

Founded in 2000, the Maurice R. Greenberg Center for Geoeconomic Studies (CGS) at the Council on Foreign Relations (CFR) works to promote a better understanding among policymakers, academic specialists, and the interested public of how economic and political forces interact to influence world affairs. Globalization is fast erasing the boundaries that have traditionally separated economics from foreign policy and national security issues. The growing integration of national economies is increasingly constraining the policy options that government leaders can consider, while government decisions are shaping the pace and course of global economic interactions. It is essential that policymakers and the public have access to rigorous analysis from an independent, nonpartisan source so that they can better comprehend our interconnected world and the foreign policy choices facing the United States and other governments.

The center pursues its aims through

- Research carried out by CFR fellows and adjunct fellows of outstanding merit and expertise in economics and foreign policy, disseminated through books, articles, and other mass media;
- Meetings in New York, Washington, DC, and other select American cities, where the world's most important economic policymakers and scholars address critical issues in a discussion or debate format, all involving direct interaction with CFR members;
- Sponsorship of roundtables and Independent Task Forces whose aims are to inform and help to set the public foreign policy agenda in areas in which an economic component is integral; and
- Training of the next generation of policymakers, who will require fluency in the workings of markets as well as the mechanics of international relations.

Council Special Reports
Published by the Council on Foreign Relations

The National Interest and the Law of the Sea
Scott G. Borgerson; CSR No. 46, May 2009

Lessons of the Financial Crisis
Benn Steil; CSR No. 45, March 2009
A Maurice R. Greenberg Center for Geoeconomic Studies Report

Global Imbalances and the Financial Crisis
Steven Dunaway; CSR No. 44, March 2009
A Maurice R. Greenberg Center for Geoeconomic Studies Report

Eurasian Energy Security
Jeffrey Mankoff; CSR No. 43, February 2009

Preparing for Sudden Change in North Korea
Paul B. Stares and Joel S. Wit; CSR No. 42, January 2009
A Center for Preventive Action Report

Averting Crisis in Ukraine
Steven Pifer; CSR No. 41, January 2009
A Center for Preventive Action Report

Congo: Securing Peace, Sustaining Progress
Anthony W. Gambino; CSR No. 40, October 2008
A Center for Preventive Action Report

Deterring State Sponsorship of Nuclear Terrorism
Michael A. Levi; CSR No. 39, September 2008

China, Space Weapons, and U.S. Security
Bruce W. MacDonald; CSR No. 38, September 2008

Sovereign Wealth and Sovereign Power: The Strategic Consequences of American Indebtedness
Brad W. Setser; CSR No. 37, September 2008
A Maurice R. Greenberg Center for Geoeconomic Studies Report

Securing Pakistan's Tribal Belt
Daniel Markey; CSR No. 36, July 2008 (Web-only release) and August 2008
A Center for Preventive Action Report

Avoiding Transfers to Torture
Ashley S. Deeks; CSR No. 35, June 2008

Global FDI Policy: Correcting a Protectionist Drift
David M. Marchick and Matthew J. Slaughter; CSR No. 34, June 2008
A Maurice R. Greenberg Center for Geoeconomic Studies Report

Dealing with Damascus: Seeking a Greater Return on U.S.-Syria Relations
Mona Yacoubian and Scott Lasensky; CSR No. 33, June 2008
A Center for Preventive Action Report

Climate Change and National Security: An Agenda for Action
Joshua W. Busby; CSR No. 32, November 2007
A Maurice R. Greenberg Center for Geoeconomic Studies Report

Planning for Post-Mugabe Zimbabwe
Michelle D. Gavin; CSR No. 31, October 2007
A Center for Preventive Action Report

The Case for Wage Insurance
Robert J. LaLonde; CSR No. 30, September 2007
A Maurice R. Greenberg Center for Geoeconomic Studies Report

Reform of the International Monetary Fund
Peter B. Kenen; CSR No. 29, May 2007
A Maurice R. Greenberg Center for Geoeconomic Studies Report

Nuclear Energy: Balancing Benefits and Risks
Charles D. Ferguson; CSR No. 28, April 2007

Nigeria: Elections and Continuing Challenges
Robert I. Rotberg; CSR No. 27, April 2007
A Center for Preventive Action Report

The Economic Logic of Illegal Immigration
Gordon H. Hanson; CSR No. 26, April 2007
A Maurice R. Greenberg Center for Geoeconomic Studies Report

The United States and the WTO Dispute Settlement System
Robert Z. Lawrence; CSR No. 25, March 2007
A Maurice R. Greenberg Center for Geoeconomic Studies Report

Bolivia on the Brink
Eduardo A. Gamarra; CSR No. 24, February 2007
A Center for Preventive Action Report

After the Surge: The Case for U.S. Military Disengagement from Iraq
Steven N. Simon; CSR No. 23, February 2007

Council Special Reports

Darfur and Beyond: What Is Needed to Prevent Mass Atrocities
Lee Feinstein; CSR No. 22, January 2007

Avoiding Conflict in the Horn of Africa: U.S. Policy Toward Ethiopia and Eritrea
Terrence Lyons; CSR No. 21, December 2006
A Center for Preventive Action Report

Living with Hugo: U.S. Policy Toward Hugo Chávez's Venezuela
Richard Lapper; CSR No. 20, November 2006
A Center for Preventive Action Report

Reforming U.S. Patent Policy: Getting the Incentives Right
Keith E. Maskus; CSR No. 19, November 2006
A Maurice R. Greenberg Center for Geoeconomic Studies Report

Foreign Investment and National Security: Getting the Balance Right
Alan P. Larson, David M. Marchick; CSR No. 18, July 2006
A Maurice R. Greenberg Center for Geoeconomic Studies Report

Challenges for a Postelection Mexico: Issues for U.S. Policy
Pamela K. Starr; CSR No. 17, June 2006 (Web-only release) and November 2006

U.S.-India Nuclear Cooperation: A Strategy for Moving Forward
Michael A. Levi and Charles D. Ferguson; CSR No. 16, June 2006

Generating Momentum for a New Era in U.S.-Turkey Relations
Steven A. Cook and Elizabeth Sherwood-Randall; CSR No. 15, June 2006

Peace in Papua: Widening a Window of Opportunity
Blair A. King; CSR No. 14, March 2006
A Center for Preventive Action Report

Neglected Defense: Mobilizing the Private Sector to Support Homeland Security
Stephen E. Flynn and Daniel B. Prieto; CSR No. 13, March 2006

Afghanistan's Uncertain Transition From Turmoil to Normalcy
Barnett R. Rubin; CSR No. 12, March 2006
A Center for Preventive Action Report

Preventing Catastrophic Nuclear Terrorism
Charles D. Ferguson; CSR No. 11, March 2006

Getting Serious About the Twin Deficits
Menzie D. Chinn; CSR No. 10, September 2005
A Maurice R. Greenberg Center for Geoeconomic Studies Report

Both Sides of the Aisle: A Call for Bipartisan Foreign Policy
Nancy E. Roman; CSR No. 9, September 2005

Forgotten Intervention? What the United States Needs to Do in the Western Balkans
Amelia Branczik and William L. Nash; CSR No. 8, June 2005
A Center for Preventive Action Report

A New Beginning: Strategies for a More Fruitful Dialogue with the Muslim World
Craig Charney and Nicole Yakatan; CSR No. 7, May 2005

Power-Sharing in Iraq
David L. Phillips; CSR No. 6, April 2005
A Center for Preventive Action Report

Giving Meaning to "Never Again": Seeking an Effective Response to the Crisis in Darfur and Beyond
Cheryl O. Igiri and Princeton N. Lyman; CSR No. 5, September 2004

Freedom, Prosperity, and Security: The G8 Partnership with Africa: Sea Island 2004 and Beyond
J. Brian Atwood, Robert S. Browne, and Princeton N. Lyman; CSR No. 4, May 2004

Addressing the HIV/AIDS Pandemic: A U.S. Global AIDS Strategy for the Long Term
Daniel M. Fox and Princeton N. Lyman; CSR No. 3, May 2004
Cosponsored with the Milbank Memorial Fund

Challenges for a Post-Election Philippines
Catharin E. Dalpino; CSR No. 2, May 2004
A Center for Preventive Action Report

Stability, Security, and Sovereignty in the Republic of Georgia
David L. Phillips; CSR No. 1, January 2004
A Center for Preventive Action Report

To purchase a printed copy, call the Brookings Institution Press: 800.537.5487.
Note: Council Special Reports are available for download from CFR's website, www.cfr.org.
For more information, email publications@cfr.org.